+REAL% WORLD %MATH=

BUILDINGS

by Paige Towler

2 4
6 8

Children's Press®
An imprint of Scholastic Inc.

Library of Congress Cataloging-in-Publication Data
Names: Towler, Paige, author.
Title: Buildings / Paige Towler.
Description: First edition. | New York : Children's Press, an
    imprint of Scholastic Inc, 2021. | Series: Real world math |
    Includes index.
Audience: Ages 5-7. | Audience: Grades K-1. | Summary: "This
    book introduces young readers to math concepts around
    buildings"— Provided by publisher.
Identifiers: LCCN 2021000039 (print) | LCCN 2021000040 (ebook)
    | ISBN 9781338762426 (library binding) | ISBN 9781338762433
    (paperback) | ISBN 9781338762440 (ebook)
Subjects: LCSH: Buildings—Mathematics—Juvenile literature. |
    Architecture—Mathematics—Juvenile literature.
Classification: LCC TH149 (print) | LCC TH149 (ebook) | DDC
    690.2—dc23
LC record available at https://lccn.loc.gov/2021000039
LC ebook record available at https://lccn.loc.gov/2021000040

10 9 8 7 6 5 4 3 2 1   22 23 24 25 26

Printed in Heshan, China 62
First edition, 2022

Series produced by WonderLab Group, LLC
Book design by Moduza Design
Photo editing by Annette Kiesow
Educational consulting by Leigh Hamilton
Copyediting by Vivian Suchman
Proofreading by Molly Reid
Indexing by Connie Binder

CONTENTS

LET'S GO!

Tall skyscrapers, strong castles, and colorful towers—how are these mighty buildings made? Let's find out! Get ready to learn about buildings all around the world. Grab a hard hat, a ruler, a notebook, and your math skills!

Math helps people make amazing buildings. The people who design and construct buildings are called **architects**. They can design using certain shapes to make structures strong. They can use **patterns** to decorate buildings. And architects use **odd and even numbers** in their designs.

Today we are traveling around the world to learn about buildings using math! Our first stop is an old city in a rainforest. Are you ready?

construction site

Katherine

MEET KATHERINE

We are joining architect Katherine Williams to learn about buildings. Katherine designs buildings and leads **construction** projects. She creates buildings that are used in different ways. Some buildings are places to live, while others are schools. Katherine decides how each building will look. She also makes sure that they have everything the people inside will need.

Deep inside a rainforest in Mexico, you'll find a 1,500-year-old city. Chichén Itzá (chee-chen eet-SAH) was built by ancestors of the **Maya** people. It is full of stone buildings, walls, and statues. As we look at the different buildings, let's get ready to find shapes.

DESIGN WITH KATHERINE

When Katherine makes buildings, she thinks about shapes. She designs some buildings to be long and tall, like a thin rectangle. Others are smaller and look like a square. She even designs some buildings with a triangle-shaped roof. That way, snow or water can easily slide off the building.

angle

One shape used in some old buildings is called a **triangle**. Triangles have three straight sides. They also have three corners, where two sides come together at one point. And they have three angles. **Angles** are the spaces formed between two lines that meet at a corner.

The most famous building in Chichén Itzá is called Kukulkan. It is a type of building known as a **pyramid**. Pyramids are buildings that are shaped like triangles. They have a wide base that makes them very sturdy. Pyramids will not fall over, even after a long time.

Kukulkan

YOU CAN DO IT! 1

What makes a pyramid special?
Its shape! Based on its shape,
which of these buildings is a pyramid?

A

tower

B

ancient tomb

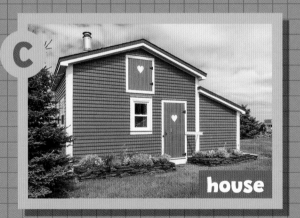

C

house

Let's move on to our next stop.

This place also has beautiful old buildings. But these
buildings are made of many different shapes!

Welcome to Thailand! In this country, there is a type of tower called a **pagoda**. A pagoda is a building that can be used as a **temple**. These buildings are tall. People come to pray at this tall pagoda that is designed to look like a mountain. The mountain is special in their religion.

DESIGN WITH KATHERINE

Katherine uses patterns when designing the outside of a building. She can arrange stones or bricks in patterns. These patterns can be based on the stones' shape, size, or color. This makes the buildings nice to look at.

patterns

People used math to build pagodas and also to decorate them.

As we look at the towers, let's search for patterns. A pattern is a type of decoration with a repeated design.

Some pagodas are painted in bright colors. Others have detailed carvings or huge statues. Some are even covered in gold! And many have pretty patterns.

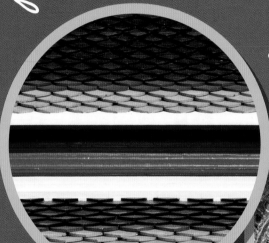

colorful patterns

detailed gold patterns

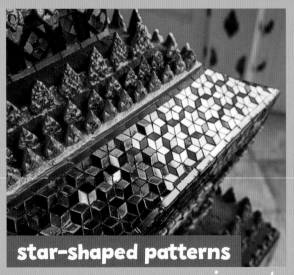

star-shaped patterns

YOU CAN DO IT!

Oops! These patterns weren't finished. Can you figure out what is missing from each pattern?

A OX+OX+OX+ _ X

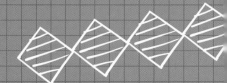

B OXO+OO+OXO+O _

C OX _ O+X _ +O+OXXO+XO+ _

Good job finishing the patterns!

Math helps some buildings look beautiful. But get ready to use math in a different way. At our next stop, the buildings not only look nice, but also keep out danger.

doors

Raise the drawbridge! Castles are large buildings. They were used across Europe for hundreds of years. Castles were both places to live and places to keep people safe. They had enough space for many people to eat and sleep. And castles had tall towers where guards could watch for danger.

rooms

windows

towers

When planning out a castle, builders placed similar items into different groups. These groups are called **categories**. Then they could count all the items in each category: rooms, towers, doors, and more. That way, the builders would know there was enough protection and space for all the people inside.

This castle has seven towers. It has one big door. And it has one heavy drawbridge that can be pulled up so that people outside cannot come in. These are all categories of parts you can find on a castle. These parts keep the people inside safe.

tower

window

door

drawbridge

YOU CAN DO IT!

③

These castles have many different parts. There are windows, doors, and towers. Each part can form a category. Can you count the items in each category? Do you see any other categories on these castles?

A

B

Terrific! Are you ready to look at more buildings? Our next stop has lots and lots of buildings—all in one spot!

 Fast trains full of people speed on underground tracks. Bright lights shine from thousands of windows. And there are so many buildings! Tokyo, Japan, is the city with the most people in the world.

DESIGN WITH KATHERINE

When Katherine plans a building, she makes sure that there will be enough space for the people inside. Katherine thinks about what each room will be used for. She thinks about what furniture the people will be using. Katherine also makes sure the rooms are big enough for people to move around safely.

Tokyo's crosswalks

There are lots of different types of buildings in big cities. Some
buildings, like homes or shops, can be small. But cities like Tokyo can be very crowded. Architects need to design big buildings that can hold many people at once. Let's get ready to **compare** types of buildings to learn more about cities.

19

Some buildings, like skyscrapers, have many stories. A lot of people can live or work in these tall buildings. Others buildings fit a lot of people inside by being long or wide.

long school building

tall skyscrapers

YOU CAN DO IT!

Let's compare sizes: Which building is the tallest? Which building is the shortest? Put the buildings in order from shortest to tallest.

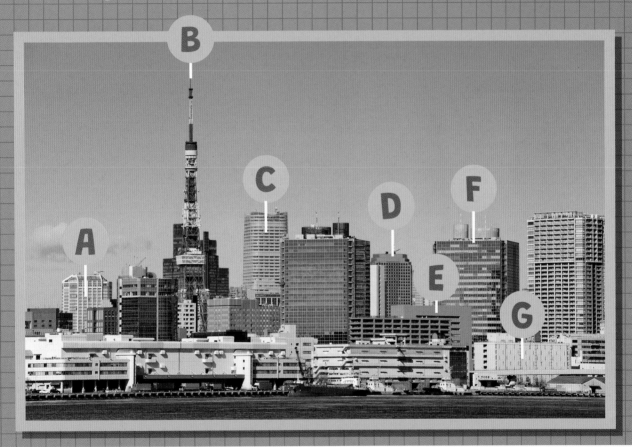

We have traveled all around the world

and looked at many different kinds of buildings. The next building could be right in your backyard!

1, 3, 5, 7, 9

What a view! Buildings can even be found among the treetops. Treehouses can be used for **shelter**. They can be used to keep things safe. Some treehouses are homes that families live in. Others store food away from animals. And some treehouses are made for playing!

2,4,6,8

Architects use math to decide how treehouses will look. They use odd and even numbers to plan out the parts of a home. Those parts include everything from the decorations to the number of floors. Even numbers can be split into two equal groups. Odd numbers cannot.

Architects need to figure out how many windows and doors a treehouse should have. Sometimes they build treehouses with other fun things!

This treehouse was built with windows. It has a door. It even has two pulleys! When you count up every part of a treehouse, the number of parts is either an odd or even number.

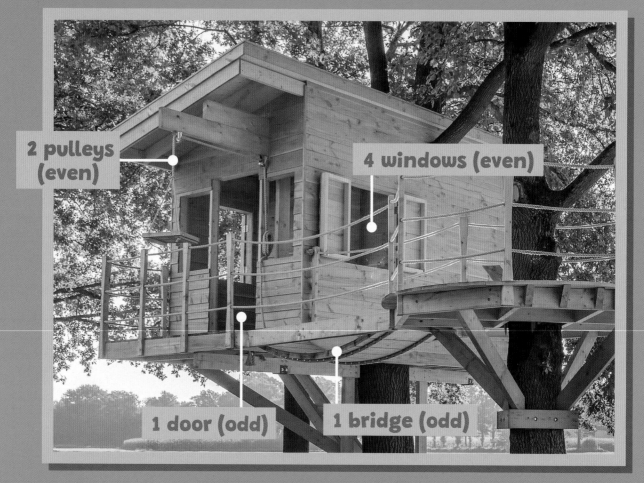

2 pulleys (even)

4 windows (even)

1 door (odd)

1 bridge (odd)

YOU CAN DO IT!

Take a look at the parts of these treehouses. Count the numbers of windows, doors, and steps on each treehouse. Are they odd or even numbers?

A

B
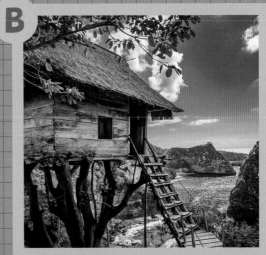

 Awesome! There is so much you can learn about buildings using math. Math is important for learning about buildings from all around the world!

WAY TO GO!

Great job exploring buildings!

We traveled around the world to learn about big buildings and small buildings, old buildings and new ones.

We also learned how architects design and plan buildings using math! We looked at shapes, patterns, categories, comparisons, and odd and even numbers. This kind of math helps architects make buildings beautiful and safe. Math is so important every day in so many ways. You might be surprised how often you use it!

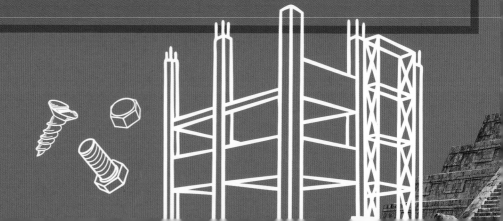

YOU CAN DO IT! 6

How many types of buildings did you see in this book? Which place had the tallest buildings?

Take a look at your own home. How many windows does it have? What shape is it?

Kukulkan

apartment building

KATHERINE DESIGNS BUILDINGS

Katherine Williams works as an architect in Virginia, United States. There, she leads construction projects for a school. Designing new buildings is a big part of her job.

Lots of people live and work at the school. There are students and teachers who use classrooms. There are scientists who experiment in laboratories. And others live in apartments or dormitories. Math helps Katherine plan how each building is going to look on the outside. Katherine can use patterns to make buildings look beautiful. She uses items from different categories, like windows or doors.

These make an apartment look bright and comfortable. Katherine can also design buildings in different shapes. These shapes help keep a building cool in the summer and warm in the winter.

building plan

Math also helps Katherine design the inside of a building. Katherine finds out what kinds of furniture people will need. Some people will need desks. Others will need beds. Katherine looks at the shape of each piece of furniture. She also measures each room. She makes sure all the furniture and people will fit. And those are just some of the math skills Katherine uses every day!

Katherine designs buildings that are wonderful places for people to live, learn, and work. Inside and outside, Katherine makes sure each building is perfect.

GLOSSARY

a home on wheels

angle (ANG-guhl): the area formed by two lines that start at the same point and go in different directions

architect (AHR-ki-tekt): someone who designs buildings and supervises the way they are built

category (KAT-uh-gor-ee): a group of people or things that have certain characteristics in common

compare (kuhm-PAIR): to judge one thing in relation to another in order to see the similarities and differences

construction (kuhn-STRUHK-shuhn): the act or process of building or constructing something

even number (ee-vuhn NUHM-bur): a number that can be divided exactly by two without leaving a remainder

Maya (MYE-uh): a member of a group of Native American tribes who live in southern Mexico and Central America

odd number (ahd NUHM-bur): a number that cannot be divided evenly by two and will always have a remainder of one

pagoda (puh-GOH-duh): a shrine or temple in some Eastern religions

pattern (PAT-urn): a repeating arrangement of colors, shapes, and figures

pyramid (PIR-uh-mid): a solid shape with a polygon as a base and triangular sides that meet at a point on top; usually has a square base and four sides

shelter (SHEL-tur): a place that offers protection from bad weather or danger

skyscraper (SKYE-skray-pur): a tall building with many stories

temple (TEM-puhl): a building used for worshipping a god or gods

triangle (TRYE-ang-guhl): a shape with three straight sides and three angles

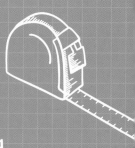

YOU CAN DO IT! ANSWER KEY

(1) PAGE 9 B: The ancient tomb is the pyramid.

(2) PAGE 13
A: OX + OX + OX + <u>OX</u>
B: OXO + OO + OXO + <u>OO</u>
C: OX<u>X</u>O + X<u>O</u> + O + OXXO + XO + <u>O</u>

(3) PAGE 17 Castle A: windows: 8, door: 1, towers: 4; Castle B: windows: 27, doors: 0, towers: 7. Other categories include flags and walls.

(4) PAGE 21 B is the tallest; G is the shortest. Shortest to tallest: G, E, A, D, F, C, B

(5) PAGE 25 A: 4 windows (even), 1 door (odd), 5 steps (odd); B: 1 window (odd), 1 door (odd), 8 steps (even)

(6) PAGE 27 There are eight types of buildings: pyramids, pagodas, castles, skyscrapers, schools, treehouses, apartments, and homes. Cities have the tallest buildings.

INDEX

Page numbers in **bold** indicate illustrations.

ABOUT THE AUTHOR

Paige Towler is a children's book author and editor. Formerly an editor for National Geographic Kids Books, she currently writes and edits for Smithsonian Licensed Publishing, National Geographic Books, Scholastic, and more. Her books include *Yoga Animals* and *Girls Can!*